WHY DO ANIMALS DO THAT?

WHY DO ANIMALS HIBERNATE?

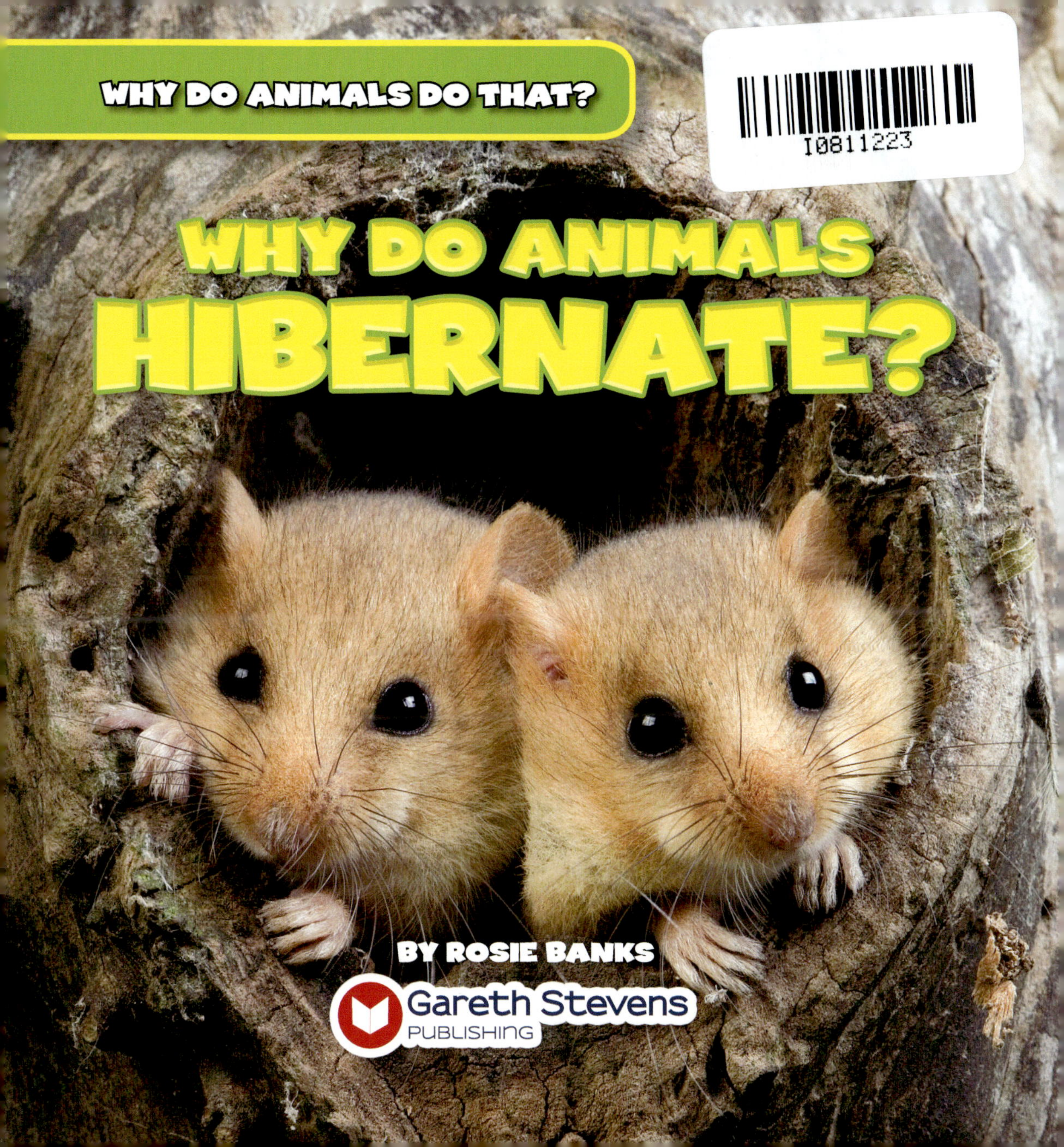

BY ROSIE BANKS

Gareth Stevens PUBLISHING

Please visit our website, www.garethstevens.com. For a free color catalog of all our high-quality books, call toll free 1-800-542-2595 or fax 1-877-542-2596.

Library of Congress Cataloging-in-Publication Data

Names: Banks, Rosie, 1978- author.
Title: Why do animals hibernate? / Rosie Banks.
Description: New York : Gareth Stevens Publishig, [2024] | Series: Why do animals do that? | Includes bibliographical references and index.
Identifiers: LCCN 2022046001 (print) | LCCN 2022046002 (ebook) | ISBN 9781538285152 (library binding) | ISBN 9781538285145 (paperback) | ISBN 9781538285169 (ebook)
Subjects: LCSH: Hibernation–Juvenile literature. | Animal behavior–Juvenile literature.
Classification: LCC QL755 .B36 2024 (print) | LCC QL755 (ebook) | DDC 591.56/5–dc23/eng/20221003
LC record available at https://lccn.loc.gov/2022046001
LC ebook record available at https://lccn.loc.gov/2022046002

Published in 2024 by
Gareth Stevens Publishing
2544 Clinton Street
Buffalo, NY 14224

Designer: Claire Wrazin
Editor: Therese Shea

Photo credits: Cover slowmotiongli/Shutterstock.com; Series Art (texture) Anna Timoshenko, p. 5 COULANGES/Shutterstock.com; p. 7 Eric Isselee/Shutterstock.com; p. 9 Liz Weber/Shutterstock.com; p. 11 cola/Shutterstock.com; p. 13 NancyS/Shutterstock.com; p. 15 Michael Benard/Shutterstock.com; p. 17 Volodymyr Burkiak/Shutterstock.com; p. 19 Nick Greaves/Shutterstock.com; p. 21 vagabond54/ Shutterstock.com

Printed in the United States of America

CPSIA compliance information: Batch #CSGS24: For further information contact Gareth Stevens, New York, New York at 1-800-542-2595.

CONTENTS

Hibernating to Survive 4
What Is Hibernation? 6
Why Hibernate? 8
Preparing for Winter 10
So Cold! 12
Brumation 14
What About Bears? 16
Hot and Dry 18
Amazing Adaptation 20
Glossary 22
For More Information 23
Index 24

Boldface words appear in the glossary.

Hibernating to Survive

Have you ever wanted to sleep through a hard day? Animals face hard times too. To get through them, some animals hibernate. This **adaptation** isn't the same as sleeping, though. Hibernating animals may look dead! Hibernation helps animals **survive**.

DORMOUSE

What Is Hibernation?

When an animal hibernates, its body uses less **energy**. Its body **temperature** drops very low. Its heart beats much slower. Its breathing slows down a lot too. Dwarf lemurs may breathe just once every 10 minutes when they hibernate!

DWARF LEMUR

Why Hibernate?

Animals need food for energy. Hibernating is a way to save energy when food is **scarce**. For example, bats eat bugs. When bugs are scarce in winter, bats **migrate** or hibernate. Little brown bats can hibernate more than six months!

LITTLE BROWN BAT

Preparing for Winter

Animals prepare for hibernation. They dig a den or find a safe place to hibernate. They also eat a lot! A groundhog may eat 1 pound (454 g) of food at a time. As animals hibernate, their bodies use the extra fat from this extra food.

GROUNDHOG

So Cold!

Animals use energy just to stay warm. Hibernating animals use less energy to stay warm. An arctic ground squirrel's body temperature drops during hibernation. It can go from 99°F (37°C) to 27°F (-3°C)! The arctic ground squirrel shakes its body to warm up when needed.

ARCTIC GROUND SQUIRREL

Brumation

Reptiles and **amphibians** are cold-blooded. Their body temperature changes with their **environment**. To survive cold weather, their body systems slow down, like in hibernation. For them, it's called brumation. They may wake to drink water. Wood frogs don't. Their bodies partly freeze!

WOOD FROG

What About Bears?

Some say bears don't hibernate! Their bodies don't change as much as some hibernating animals' bodies do. Bears may move around too. Their deep sleep is called torpor. But bears go through this to save energy during winter, just like hibernating animals do.

Hot and Dry

Winter isn't the only time for animals to save energy. Some animals hibernate after fire harms their environment. Some animals hibernate during really hot and dry weather. This is called estivation. Crocodiles may estivate to escape great heat.

CROCODILE

Amazing Adaptation

Many more animals hibernate. The common poorwill is the only known hibernating bird. Bumblebee queens hibernate over winter. Some snails cover themselves in slime before they hibernate. There's much more to learn about animals and this cool adaptation!

HIBERNATION

- Body temperature lowers
- Breathing slows
- Heart rate slows

Body uses less energy

common poorwill

GLOSSARY

adaptation: A change in a type of animal that makes it better able to live in its surroundings.

amphibian: An animal that spends time on land but must have babies and grow into an adult in water.

energy: Power to make something work.

environment: The conditions that surround a living thing and affect the way it lives. Also, the natural world in which a plant or animal lives.

migrate: To move to warmer or colder places for a season.

reptile: An animal covered with scales or plates that breathes air, has a backbone, and lays eggs, such as a turtle, snake, lizard, or crocodile.

scarce: A very small amount.

survive: To live through something.

temperature: How hot or cold something is.

FOR MORE INFORMATION

BOOKS

Jaycox, Jaclyn. *Hibernation.* North Mankato, MN: Pebble, 2020.

Rogers, Marie. *Why Animals Hibernate.* New York, NY: PowerKids Press, 2021.

Stewart, Melissa. *Summertime Sleepers: Animals That Estivate.* Watertown, MA: Charlesbridge, 2021.

WEBSITES

Not Just Sleep: All About Hibernation
www.science.org.au/curious/hibernation
Learn much more about the science behind this adaptation.

10 Fascinating Facts About Hibernation
www.natgeokids.com/uk/discover/science/general-science/10-fascinating-facts-about-hibernation/
Read these facts, and share what you learn!

Publisher's note to educators and parents: Our editors have carefully reviewed these websites to ensure that they are suitable for students. Many websites change frequently, however, and we cannot guarantee that a site's future contents will continue to meet our high standards of quality and educational value. Be advised that students should be closely supervised whenever they access the internet.

INDEX

amphibians, 14

arctic ground squirrels, 12

bears, 16

birds, 20

breathing, 6, 21

bumblebees, 20

crocodiles, 18

dwarf lemurs, 6

fire, 18

food, 8, 10

heart, 6, 21

little brown bats, 8

reptiles, 14

snails, 20

temperature, 6, 12, 14, 21

weather, 14, 18

wood frogs, 14